# Baby Loves Numbers

© CREATIVE EDUCATION

## Prologue

This Giant Illustrated Colouring Book helps babies trace and colour numbers at their own pace. The book develops fine motor skills like tracing, counting, number recognition, concentration and so much more. The dexterity and numerical skills developed with this book, prepares the child for school or online learning as the case may be. This is a parent's guide to confident home-schooling.

- Creative Education

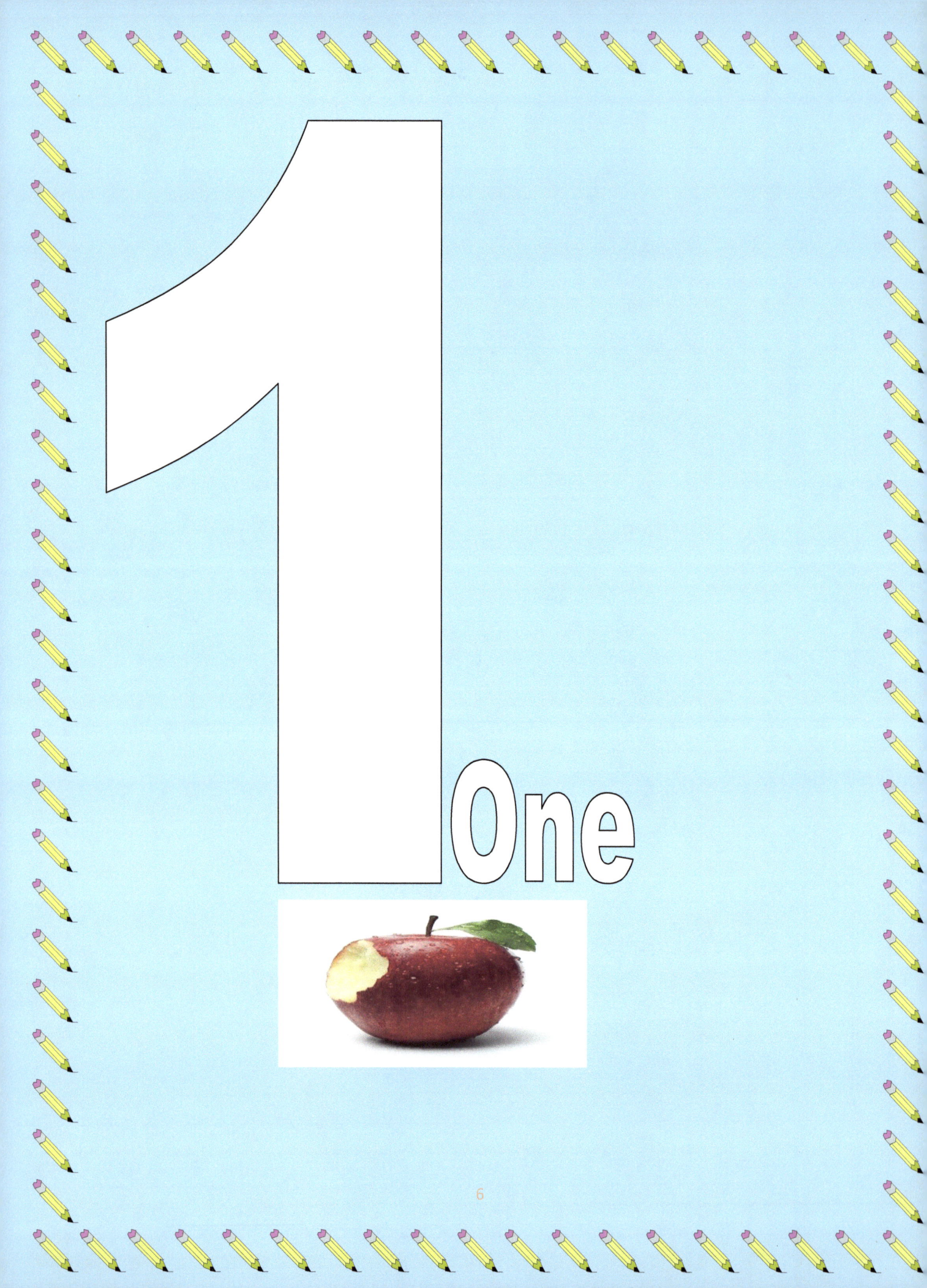

# 3 Three

# 4
## Four

# 6 Six

# 7 Seven

# 8

## Eight

# 10 TEN
# 20 TWENTY
# 30 THIRTY
# 40 FORTY
# 50 FIFTY

# 60 SIXTY

# 70 SEVENTY

# 80 EIGHTY

# 90 NINETY

# 100 HUNDRED

**THE END**

# CREATIVE EDUCATION

Creative Education strives to make learning fun for children. All the books are designed by qualified teachers and ensure that your child is prepared for school in the most creative way. The books impart basic skills that a child of the age range 2-7yrs is expected to possess.

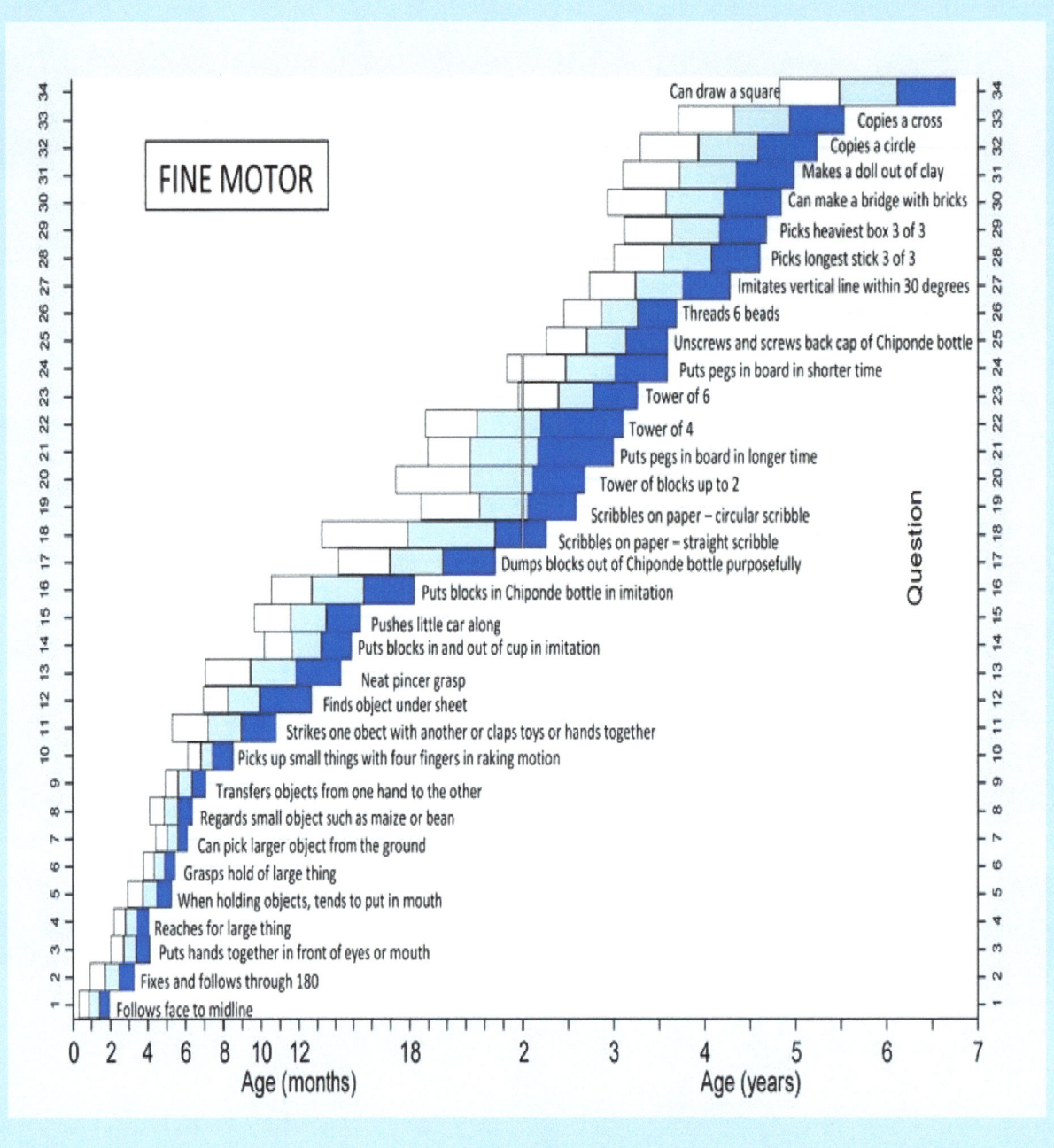

**More books in the series**

*Baby Loves Alphabets*

*Baby Loves Shapes*

*Baby Loves Colours*

www.ingramcontent.com/pod-product-compliance
Lightning Source LLC
Chambersburg PA
CBHW051837210526
45473CB00005B/1908